The Pets Speak

Terri Diener

WingSpan Press

Printed in the United States of America

Published by WingSpan Press, Livermore, CA
www.wingspanpress.com

EAN 1-978-59594-117-6
ISBN 1-59594-117-7

First edition 2007

Library of Congress Control Number 2006939196

This book is dedicated to all the animals
who have shared my life

Inky, Christy, Rusty, Buffy, Salt, Jody,
Victor, Nicky, Cleo and Mayan

And to all the animals who have touched my life

And to all the humans who love me and support my life

"Lots of people talk to animals," said Pooh.
"Maybe, but ..."
"Not very many listen, though," he said.
"That's the problem," he added.

From: A.A. Milne
The House at Pooh Corner

Contents

We need another and a wiser and perhaps a more
mystical concept of animals.
Remote from universal nature, and living by
complicated artifice, man in civilization surveys the
creature through the glass of his knowledge
and sees thereby a feather magnified and the whole
image in distortion.
We patronize them for their incompleteness, for
their tragic fate of having taken
form so far below ourselves.

And therein we err, and greatly err.
For the animal shall not be measured by man.

In a world older and more complete than ours they
move finished and complete,
gifted with extensions of the senses we have lost
or never attained, living
by voices we shall never hear.
They are not brethren, they are not underlings;
they are other nations, caught with ourselves in
the net of life and time, fellow prisoners of the
splendour and travail of the earth.

From: Henry Beston
The Outermost House

Note: an 18"x24" poster of this passage,
suitable for framing, is available from
Harold Berliner, printer
Berliner@nccn.net

Introduction

I want to invite you into my world. On a daily basis, I have the privilege of connecting intimately with animals about what they do and why they do it. It is a wonderfully honest exchange; animals are not capable of lying, cheating, or deceiving.

Once we accept that animals are thinking, feeling, sentient beings who have a lot to teach us about living, dying, relating, and forgiving, we can no longer treat them as anything less. They are different from us—not less than us.

All too often we think of them as "people in fur," but they are not. Each teaches us who they are and—if we allow it—who we are.

This book is about the insight that animals have for us. Each of the stories is unique as each animal is unique. The animals you will meet in this book are telling their story, and I chose them because they speak for every dog, every cat, every bird, every horse. They offer us the wisdom of their species.

Scientific studies have suggested evidence for telepathic communication. Mental telepathy is the transfer of thoughts, perceptions, ideas, and feelings from one mind to another. There needs to be a channel for this to occur.

There is a phenomenon in physics known as the Schumann Resonance, which states that, in the earth's atmosphere (the cavity between the surface of the earth and the ionosphere fifty-five kilometers up), there are extremely low frequency (ELF) electromagnetic waves that correspond to the rhythm of the human brain in altered states. Telepathic communication utilizes those frequencies as the highway system for sending thoughts and feelings over distance.

The ability to communicate with animals is a natural process. I believe it is encoded in our DNA; it is how we communicated when we were cave people. We had to have more communication than a few grunts and hand signals to have survived and evolved into who we are today. The telepathic capacity was put on the back burner when we developed the ability to speak.

Small children have this telepathic capacity available to them, and it is often expressed through imaginary friends, reporting what the family dog is saying, talking to trees, etc. Well-meaning (and perhaps not-so-well-meaning) adults often tell them they have a great imagination, or tell them it's not possible for trees to talk, or ridicule them for making up stories. Children believe their adult authorities, and, as a result, often block out this natural way of being. Some don't bury it, but most do.

The good news is that the capacity to communicate telepathically is a part of us, and we can reclaim it anytime we choose. For some of us, regaining our telepathic capacity comes easily and naturally; for others, it takes concentration, focus, and technique. This capacity is our birthright. With desire, attention, and practice, we can all talk to animals and hear pets speak.

The Pets Speak

No, I'm Not Dr. Dolittle

. . . and no, animals rarely speak to me in words. Animals have strong opinions and thoughts about who they are and who we are. They convey their ideas and emotions through thought-forms, mental images, feelings, and symbols. We can receive them as sensations, an intuitive "knowing," a mental picture, a thought that pops into our heads, and sometimes words.

My journey to learning this began many years ago, when I got my first cat. It was truly a pivotal moment in my life. I was moving into my own apartment and asked my vet what kind of dog I should get, given that I would be living alone and working weird hours.

He said, "Get a cat."

"No," I responded, "I don't like cats. What kind of dog should I get?"

Again he answered, "Get a cat."

After a couple more of these exchanges I told him I'd consider it.

The phone rang. It was my cousin Marcie. "Just want to make sure you're home. I want to bring you your new apartment present."

Half an hour later, in she came with a plain old cardboard box and a bag. She put the packages down and opened the box. There, huddled in a corner, was the most precious orange tabby cat, green eyes wide with curiosity and apprehension, with a note on his collar: "I'm Buffy. Feed me. I'm yours." Then she opened the bag and pulled out a bag of kitty litter and a box of Tender Vittles.

I didn't know until years later the significance of that event.

Until that point, I didn't like cats. We'd had dogs from when I was eight until I was a sophomore in college. Cats were too aloof, too independent, too unsocial—so my mother had told me. I believed it.

Buffy's first lesson for me was to debunk that belief. He taught me about wild abandonment as he raced through the apartment, launching himself over anyone or anything in the way, running through the living room to the dining room to the kitchen, back to the living room and then launching himself onto the drapes, catching hold with his claws and happily swinging away. He taught me about finding joy in little things as he splashed in the running faucet or played catch with a toy mouse. He taught me it never hurts to ask for what you want when he attacked my toes as I dressed for work, and he was not ready for a nap. He taught me about curiosity when he stuck his head under a closet door. He taught me about sharing when he welcomed—yes, welcomed—companion cats into the household. He taught me about jealousy when he peed on my couch immediately after a date left.

He taught me a lot about love and intimacy. He showed me how simple it is to receive affection and be comforted, to ask for it when wanted, and to go on to the next thing when filled up.

Sixteen years later, he also taught me about dignity in dying. That day was one of those hot, muggy days that mark summer in Maryland. Buffy had been declining for several months. He had been at the vet's for a week fighting kidney failure, and the vet suggested I take him home for the weekend and see how he did. Here was a cat who could hardly walk, and he made his way down three flights of stairs, spending time in each room he had claimed. By Monday it was clear that he was ready. He looked at me in that way that says, "It's time now."

I was with him through the whole process; that was something I never thought I could do. But how could I leave him to strangers? He was my longest roommate, one of the most intimate relationships of my life, and I loved him. As the vet performed the procedure, I wrapped my arms around him and wrapped his spirit in clear light and said goodbye as he soared out of his body into the company of a group of souls waiting to greet him.

I learned then that it's okay to grieve for non-human things. Our culture doesn't support grief. Our workplaces give people three days or a week when a spouse, parent, or sibling dies. We admire people who don't show their feelings and plunge ahead. Showing these feelings is considered a weakness in our society. But grief is one of the most profound experiences we can have. It honors the being who has died; it honors the relationship; it honors ourselves, to fully feel. It can be a time of profound creativity and connection, if we allow it.

About two hours after Buffy died, I was taking a walk and I felt his presence. It was unmistakable. "How do you feel?" I asked. "Free," I heard him say.

"Why are you hanging around? You always wanted to be free. Now you are. Go, be free," I thought to him. "I don't want to leave you yet," was his response.

That was the first time I remember getting such a clear message from an other-than-human being. What a powerful one it was. It was the beginning of consciously trusting love.

Buffy was never happy being an only cat, and he had a total of three companions in his life. First was Salt, a black and white guy who literally could not find his way out of a paper bag. Salt lived only a few months; he had swallowed the rubber tip of a doorstop, and by the time he was correctly diagnosed he had developed peritonitis. Then there was Jody, a gray and white, neurotic little cat who was terrified of just about everything. He was about two when he developed a serious kidney infection that resisted treatment.

And then came Victor—and the amplification of my learning. He joined our home when Buffy was three. They were close buddies for the next thirteen years. Victor was a handsome, sleek fellow with a beautiful coat of long black and copper fur, and a white blaze framing his face and continuing down his chest.

Victor picked me. I was ready for another cat to keep Buffy company. A vet referred me to a local person who had a cat rescue service. I decided, if one picks me out, that's the one I'll take. I was hoping for a slightly older guy who was past the hyper-kitten crazies.

Victor came up to me as soon as I sat down, looked at me confidently as if to say, "Well . . . what kept you?" and was "Velcro-cat" for the next hour. I took him home.

Buffy and Victor were my companions during a very dark time in my life. I had resigned from a secure job to start a consulting practice, and after six months my client base dried up. I moved into a working-together-living-together experiment with friends, and it was falling apart. On top of that, a major family drama was unfolding.

These two cats offered consistency and support when I was feeling isolated and disconnected. I learned then just how deep a relationship could be.

After Buffy died, I knew intuitively that Victor needed the experience of being the only and then the oldest. He took on the responsibility of being my sole protector with intense dedication to the job.

He would come running down three flights of stairs as I unlocked the front door, skid to a stop in front of me, and stare at me until I picked him up. Then he would chitter away, telling me, I assumed, about what went on while I was gone. He would camp out on the second step in the house, hissing and spitting at anyone who even thought about intruding on my living space. He read my moods and knew just when to nestle against me for silent communion.

I loved this cat deeply. He touched my soul in a way that words can't describe. I watched him grow into his role and became increasingly sensitive to how intelligent and complex he was. I had the idea then that Buffy was this way, too, but I couldn't fully see or appreciate it while he was alive.

About six months after Buffy died, a friend asked if I wanted a kitten from her father's cat's litter. Everything inside me yelled "NO" as I found myself saying, "Okay."

As an adult, I have increasingly trusted my unconscious mind, and when words "fall out of my mouth," I take it as my truth. So, I brought Nicky home.

Nicky was a roly-poly gray tabby with a little tick of a tail (he was part manx, I guess) and the sweetest face and softest eyes I had ever encountered in an animal. His behavior remarkably paralleled Jody's (remember Buffy's second companion?). Jody had been afraid of dogs since he was about three months old, when a friend brought her puppy to visit. Jody was terrified at the dog's play posturing, and showed his distress with a trembling "whooooooo" sound and a dash under the sofa.

Nicky was afraid of dogs the first time he met one. In my then-living situation, my housemates had a little silver-gray-white shih tzu named Chi. Chi only wanted to play. He and Buffy had been good buddies. Victor had dismissed him long ago, and Chi had been without a playmate since Buffy died. Nicky was afraid of Chi, and the way he whined and meowed his distress was so uncannily similar to how Jody reacted that I had my first knowing that reincarnation is a fact.

My housemates and I set about desensitizing Nicky. It took a few weeks of holding Nicky in my lap, inviting Chi over, and then removing Nick just as I felt him tense up, but before he meowed his fear. Within a short time, they became fast friends, chasing each other up and down hallways, play-ambushing each other, colluding in snatching forbidden treasures around the house.

When I moved out of that living experiment, it was Victor, Nicky, and me back in my condo. Life was harmonious for a long time, and then, suddenly, Victor started pooping outside of his litter box. First I was upset that he might be sick. Nothing conclusive showed up. At my vet's suggestion, I tried enzyme cleaners, sprays, blocking rooms off. Nothing worked. I was getting increasingly enraged. I'd wake up each morning looking for evidence of his transgression. When I found it, I would yell at him and chase him around the apartment. He eluded me. I don't know what I would have done if I had caught him.

I felt like I was in the middle of an abusive relationship and didn't know how to make it better. I called my vet regularly, desperate for more ideas.

"We've tried everything with him," she said. "Would you consider an animal communicator?"

"I don't know," I said. "Maybe. What do they do?"

"She might be able to find out why he's doing it," the vet responded.

I didn't know such people existed, but the idea fascinated me. That day I called Marlene Sandler, the communicator my vet suggested.

I sent her Victor's photo. At that time, her schedule was booked, and I had to wait an interminable two weeks for the appointment.

Marlene lived about two hundred miles away, so the whole exchange happened by phone. I called her at the appointed time. After a few minutes of introduction, Marlene said to me, "Please don't be offended by this question, but do you color your hair?"

Flabbergasted, incredulous, and irritated, I answered, "Yes."

"Do you color it at home?" she asked.

"No," I responded. "I have it done at a salon."

"Well, Victor is allergic to a hair product—it looks like a dye—and it's leaving him with a chronic headache and a mild depression," she offered.

I looked at my shampoo and, to my amazement, found that it contained red dye. I read the ingredients to Marlene, and she told me which ones were affecting him.

The litter boxes were in the bathroom. He couldn't escape it. He had a few other complaints, but that was the big one.

Marlene told me I could cleanse the space with coffee grounds and fresh mint (a superb environmental detoxifier that absorbs toxic energy just as baking soda absorbs odors). Shortly after our conversation, off I went to the grocery store in search of fresh mint. I brought it home, put it in a bowl with coffee grounds from the morning pot, and put it on a shelf in the bathroom.

Within three hours, the pall of depression had lifted from Victor, and he pooped in his box.

I called Marlene the next day and said, "I don't know what exactly you do, but I want to do it, too."

By "coincidence" (if you believe in such things), she was scheduled to teach a basic animal communication workshop in my area a few weeks later. The workshop involved doing a variety of exercises in the morning to expand our abilities to focus, send thoughts and ideas telepathically, and receive information from each other. The afternoon was devoted to practicing with the resident animals and with photos that the participants brought.

In the morning, I saw just how much I got in my own way. For example, we worked in pairs, one person sending and the other receiving. Marlene told the senders to send a shade of blue in whatever shape they chose. I received a rectangular-shaped baby blue, and then immediately embellished it to electric blue with sparkly flashes exploding in it, cascading like a waterfall. It was beautiful. It just wasn't what my partner sent. She had sent the image of a baby blue blanket.

In the afternoon, after a few hesitant starts, I began to get impressions, words, and images from the animals. And they were accurate, according to the feedback from their people. Well—you could have picked me up off the floor.

I was assigned to one of the resident dogs, a five-year-old Heinz 57 variety. As I concentrated and called him silently, I felt like I was being pushed away, got a feeling of exasperation, and then the thought "wait for me" popped into my head. A slew of my own thoughts then popped into my head: Was he talking to me? What did he want me to wait for? Why did he want me to wait? Then this little guy jumped off the deck and took off after his "siblings" who were running up the street. Apparently his message was for them, not for me. He'd had enough of these novices trying to talk to him. When I reported this exchange to his person, she laughingly said, "Yep, sounds like him." I was amazed that I had heard his thoughts so clearly.

Later on, I was working with the photo of a sixteen-year-old black cat. The only things I knew about him were his name, age, and gender. First I felt sadness, and then the thought "I'm here as long as I'm here" came into my mind. His person confirmed that he had cancer. I then got more thoughts from the cat: He knew he wouldn't recover, he knew how hard it was for his person to deal with this, and he didn't want his person to know that he was close to dying. I told his person that the cat knew he was sick, and to enjoy their time together, no matter how long or short it might be.

I drove home feeling spacey and happy. Spending a whole day in an altered state will do that. I also felt blips of excitement and possibility. I took the workshop out of curiosity and because it felt right. I wanted to enhance my relationship with my own animals. I had no idea I'd be good at this communication thing. Yet it was so easy and natural.

When I got home, I began practicing with the animals of friends, and was getting information that I didn't previously know about them. One of my friends asked me why her cat constantly followed her around the house, even though she had rescued him eight years before. "She's my buddy, and buddies stay together," was the thought that came to me.

His person's reaction? Very quietly she said, "I tell him he's my buddy all the time, even though I named him Winston."

Glimmers of possibilities started seeping in. I had a successful consulting practice in facilitating people skills in the workplace: communication, conflict resolution, group dynamics, problem solving, stress management, team building, etc. What if I could use my talents with animals and their families? After all, I'm a skilled listener—both to what is being said and to what's not being said.

Wouldn't it be wonderful if animals could make their thoughts and feelings known to their people?

Wouldn't it be exciting to help people understand that animals have definite thoughts and opinions about themselves, their lives, and us?

Wouldn't it be great to help people and their animals negotiate and create more harmony in the household?

Wouldn't it be a gift to humans and animals for people to know that animals are thinking, feeling, consciously aware beings who have a lot to teach us about living, dying, relating, and forgiving?

A few months later I took an advanced animal communication class that Marlene taught. It offered even more insight into ways to enhance the relationship between animals and their human companions.

We were working in small groups. Each person would present a situation with an animal, and the rest of the group would tune in and do a mini-session with the presenter. One person, whom I'll call Jen, had brought her six-year-old bijon friese, Hobo, and wanted to know why he would sometimes listen to her, and sometimes not.

Hobo was wise beyond his appearance. His scraggly white fur was in dire need of a bath and a trim. His little body would quiver with excitement over things that often he was the only one privy to. A feeling of joy emanated from him.

As I concentrated and tuned in to him, hoping I would make some connection, I felt what I can only describe as exasperation, and then a clear thought came into my mind: "right relationship." I waited to see if I would get anything else and then got a sense of knowing a whole story: One of his jobs was to teach Jen the right relationship between dogs and humans, and she wasn't getting it. He would do as she asked when she presented it the "right" way, and would balk when she did it "wrong." I had no idea what his ideas of right way and wrong way were, but I told her anyway.

At a break, Jen invited him to see something on the property that she wanted to see. With a singsong and tentative voice, she asked, "Hobo, would you like to go see the pond?" I could tell from her demeanor and her voice that she was frustrated and afraid that he wasn't going to respond and that he would do what he wanted to do. Sure enough, Hobo sat and wouldn't budge. She tugged on his leash. She pleaded with him to "Please come." Frustrated, Jen said to me, "See, this is what I mean."

"Do you always ask him like that?" I asked.

"Yes," she said.

"Well, he says you're doing it wrong, so tell him differently," I offered. I had no idea what that would be, however.

Then she tried telling him, "Hobo, come. Let's go to the pond." This time, her demeanor was confident, and her voice was strong. She knew what she wanted him to do, and firmly conveyed the message. Hobo's response? He picked himself up and trotted along beside her.

This was a textbook example of what I was discovering about communicating with animals. They respond first to our feelings, then to the picture of our thoughts, and finally to the words we use. They want to please us. However, if we tell them something while thinking or feeling something else, they will respond to the feelings and thoughts. With Hobo, Jen was feeling anxious; she was picturing him sitting, and the tone of voice she used with him was tentative. Once she shifted to a feeling of certainty, with the thought that he would do whatever she asked him to, she pictured him going with her. Her voice became firm and confident. And Hobo responded in a canine version of "Okay, she's got it."

As with the first workshop that I attended, I left this one with a deep sense of comfort and familiarity that amazed me. This felt so natural. The information just flowed once I relaxed and allowed the connection to occur, instead of trying to make something happen.

A few months after that, Marlene called and asked me if I would work with a lost animal. A bit daunted, I said, "Okay."

Thus it began. It didn't take long for word of mouth to bring a slew of lost animal cases. Then calls came about behavior problems, relationship issues, and health and illness crises. I discovered that I was not so good at finding lost animals. However, I could connect quickly with the emotional, spiritual, and physical experiences of animals, and could bridge the communication gap between their people and them, even if the animal had died.

I was still dabbling in this work when Victor went into a downward spiral. He had a mass in his intestines that everyone was pretty certain was malignant. I chose to not pursue a formal diagnosis because I would not subject him to surgery or chemo. He was sixteen, and I got the sense that he wanted nothing intrusive. I felt his life force diminish almost daily. I knew he was ready to go, and wanted to honor his wishes, but I didn't know how to give him permission to go. A part of me wanted him to be with me forever, and I didn't yet trust the connection we can have with someone who has died.

I began a life review with him, which I had read about in the process of educating myself for this work. It was an intense, touching, solidifying experience as I reminisced with him about our life together. I would become aware of how much I loved him, then hold him and stroke him as I remembered the funny, touching, and frustrating events. Over the next few weeks, he would come over to me at all kinds of odd times and stare at me in his way, until I picked him up. He would even do this if company were visiting, which he never would have done before.

A thought popped into my head that he wanted another cat to hold the space until he came back, because he was planning to do just that—reincarnate as a cat again and find me. I was astounded and humbled—first, that I got such a clear message from him when I was so vulnerable; second, that his wish was so clear; third, that he was planning on coming back; and fourth, that he wanted to come back to ME.

I was still having trouble trusting myself on this, so I called Marlene for verification. She was at a retreat. I had to depend on myself on this one. It was a little test, I believe.

A friend and I went out to a rescue service, Cat Woman Cat Rescue, which was housed in a four-bedroom rancher occupied by two adults and thirty-odd cats. Cats were everywhere. Some were hugging a wall to escape the chaos. Some were climbing over the rescuers. Some were fighting. Some were adoptable and some were permanent residents because of physical or behavioral problems. The Cat Woman pointed out the fifteen available for adoption. Whom should I pick? This was an important decision, and I felt like I was in outer space.

I selected a female orange tabby who was hanging out in the corner of the hallway. I felt a kindred spirit with her. We both got overwhelmed with frantic activity, and I can easily hide in a corner when it gets too chaotic. I got the thought, "Get me out of here." I knew it wasn't my thought.

Next came the test of getting Victor and Nicky's agreement. Calling on everything I knew about communicating with animals, I concentrated and focused on Victor and Nicky, sent them a mental picture of this orange striped girl, and tried to embody the feel of her personality, energy, and demeanor. I wasn't sure what I got back from them, but it felt okay. More accurately, I didn't get a screaming "NO" in my head.

The Rescue people had named her Elizabeth, but I didn't like that name and she didn't respond to it. She just didn't feel like an Elizabeth.

Once we got home, I tried out several names. I liked Maya, but she didn't respond to that either, and it felt too intense for her. Then, totally unexpectedly, the name "Cleo" popped into my brain. When I called her "Cleo," she stopped in mid-track and just looked at me. I felt an acknowledgment. So, Cleo it was. I was becoming sensitized to the fact that animals have clear opinions about their names.

I set her up in my office with the door open and her own litter box, food, and water. I planned for her to be there as long as she needed to be, and would leave the timing of her merging in the household up to her. Victor, however, had other ideas. Remember now, he's the one who wanted another cat to take care of me until he returned.

Victor came into the room and stared her down. He used her litter box. He ate her food. After a day of this, I'd had enough.

Victor was lying on the dining room table—one of his favorite places—in casually alert mode, taking in everything that was happening. His eyes were narrowed, his brow furrowed, his ears back. I could feel his anger. His energy was stronger than it had been in weeks.

I sat at the table and said out loud to him, "You can't have it both ways, Victor. I brought her here because you said you wanted it. She can't be here and not be here at the same time. So, you decide when she is acceptable and let me know."

Nearly immediately, the layer of anger that had settled over him lifted. He felt lighter, his ears perked up, and his face smoothed out. The next day, when I returned home, I found them sleeping peacefully near each other on the office sofa.

At that point, Marlene returned from her retreat and called me back. She confirmed it all and told me that Victor was ready any time I was. He could last another few months, but he didn't want me to see him in his weakness. He wanted music and candles for the experience. He also wanted me to throw a party afterwards, with all the people he liked, and I would know who they were.

I was really surprised at that. I didn't think he liked enough people to invite to a party. He was right, though. As

I thought of each person who knew him, I would get a warm, settled, contented feel about some and a cool disconnect about others. By the end of the week, when the vet was scheduled to come to my house to euthanize him, I knew whom to invite. He had sent the guest list to me.

That morning, Marlene called. She had a message from Victor. "He wants you to know how grateful he is that you let him be who he is, and helped him find his way. You gave him room to work out his fears," she said.

Barely able to speak, I responded, "That's what you do when you love someone."

"He says you and he share that issue, and he was so happy to be part of your getting it."

He was feisty to the end. He didn't exactly fight the vet. He hissed at her and batted her a few times, then put his head on my shoulder as he drifted out of his body.

I knew that animal souls take time to rest, rejuvenate, and review their life experiences, just as people do. Marlene had told me long before that animals usually take about four to six weeks to decide if they are coming back, and when.

Exactly four weeks later, Marlene and I checked in with him. We both tuned in in our accustomed way—focusing, visualizing him, and calling him three times—and there he was. I could feel his presence, but couldn't get any of his thoughts. Marlene started laughing. "He says he's ready to come back, but they don't have their act together here on the continuum. They lost his number and made him go to the back of the line."

Probing a little further, she told me Victor's guide had come in and said he needed "karmic reduction and molecular restructuring," whatever that meant.

Indeed, Victor did do much healing. He would pop into my awareness, or I would dream about him regularly. I could sense he was strong and happy.

About two years later, he came into one of my meditations and said, "I'm coming back, and this is what I'm going to look like. Find me. You'll know me by my eyes." The image he projected to me was of himself as I knew him, and of himself as a kitten with charcoal gray fur and black stripes and mysterious green eyes. I tried out a few names, and he liked "Mayan" the best.

A few months after that, I knew he was back. When I tuned in, he felt more solid than in our preceding visits. Not trusting myself—again (it's difficult to get information when one is emotionally connected to the message)—I called Marlene. She confirmed that he had been born.

A few weeks after that, I felt compelled to go to the local Humane Society. I told Nicky and Cleo that Victor was back and I was going to get him. But he wasn't there. There were no charcoal-gray-with-black-stripes kittens available. My insecurities skyrocketed. I felt him. I just knew I would find him there. Was I fooling myself? Is this stuff real? Should I retire my shingle?

Half a week later, I felt I had to go back to that Humane Society.

And there—in a cage that was exactly at my eye level—were two charcoal-gray-with-black-stripes kittens. They were brothers, part of a litter that had been picked up by animal control. This shelter is no-kill, and they allowed the mom to wean the kittens. These two guys were eight weeks old.

Which one do I pick? One was identified as #354896 C and the other #354896 D. C was slightly closer to me, so I picked him up first. He made himself very heavy in my hands (as heavy as an eight-week kitten can be), and made eye contact. I got chills running up and down my back. I picked up his brother, whose only connection was to use me as a springboard to pounce on a stuffed mouse.

I filled out the application on the spot, and took C home two days later. He answered to the name Mayan within hours.

The learning goes on. I think it will be continuous. As I watch my own animals grow and work out their issues (for they have issues just as we do), and I understand from them why they are here—what their purpose is in this life—I feel humbled and grateful. From them I have learned how to connect to other people's animal companions. They are my guides and helpers. Sometimes I ask them to go in and find out what is going on with an animal client.

To be allowed in—to know the essence of another being so intimately, to be able to recognize each one's uniqueness, to understand another species, to truly appreciate who they are, to bridge the communication gap between the species—and to be able to teach other people to do all of this is very precious. I am honored to be their voice as the pets speak.

A Typical Session

The idea of animals being able to communicate their thoughts and feelings is a stretch for many people. They accuse us of anthropomorphizing our animal companions. That's simply not how it is. Sometimes the best way to get people to understand what's going on with their animals is through analogies to human behavior and attitudes.

That doesn't mean the animals are people in fur or with wings. True connection happens when we understand an animal's actions in terms of what is natural and instinctive for that species

Communicators get messages in different ways. Some of us see mental pictures, some hear words, and some get a "felt sense"—feelings and thoughts that we know are not ours—that come in when we are in a relaxed, meditative state.

I prefer working by phone. On-site visits take longer, and there is the distraction factor. Animals want to know who you are, are you there to play with them, etc. By phone, once I tune in, I often get some thought-form about "Who are you and how did you get into my head?" Once they realize I'm in their heads, and not in their houses, they usually drop their guardedness.

I begin with the person on the phone and a photo of the animal. I chat briefly with the person to find out what the situation is, and what they would like to accomplish in the session. Some people have a long list of questions, and I like to get the whole picture of their concerns before tuning in to their animal companion. I explain to them that this will be like a three-way conference call, with one of the parties communicating silently. My job is to be as clear an interpreter as I can be for both of them.

I then ask the person to get relaxed and comfortable, to take a few deep breaths, and to shift over to an open, relaxed, meditative state. This is a natural state for all human beings, It is not, however, the one that we are in all of the time, so we need to consciously shift over to it.

I do the same thing. I also initiate a short ritual. Knowing that we all have invisible "helpers," I call in my guides, as well as those of the animal and the person, and ask that they assist with this session. In addition, I silently state, "Only those energies that are serving the highest good of all may stay present."

Once I have shifted over to this altered state, I ask the person to visualize the animal and mentally call him or her silently three times. There is nothing magical about three times. It is a quick way to concentrate and focus. I also focus and call the animal. In the early days, I would send out a feeling of connection from my heart to the animal's heart, and send out a "scanner beam" from the middle of my forehead until I clicked in to a mental connection. Now, it's automatic. As both the person and I visualize the animal, I feel tuned in.

The animal does not have to be in the room with his/her person. Because it is a telepathic connection, the person could be in Italy and the animal in Maryland, and the link would be just as strong.

I get immediate impressions as we make this connection. These are usually about the essence of this animal, along with what s/he is feeling. I report these to the person. Sometimes animals are very chatty, and sometimes I get a thought of "Oh, hi—what do you want?"

All of this comes to me in the form of thoughts and feelings. I rarely see pictures or hear actual words. When I tell the person that Fluffy said such and such, the message is coming from thoughts that pop into my head. Because I have centered myself and shifted over to an altered state, I know that my impression slate is clear, and that whatever comes into it is the message from the animal. If I am in an emotional or mental state where I can't be clear, I reschedule the session.

Animals deliver their message very quickly. We either get it or we don't. In a half-hour session, I often spend about five minutes talking directly with the animal, and the remaining twenty-five helping the people understand why this animal is behaving this way, and develop strategies to resolve the situation.

More often than not, an animal's behavior or illness reflects something going on with the person or the household. They show us how our reactions and thoughts affect us physically and emotionally. Most of the time, people are open to this information. In fact, most people quickly see the parallels between what the animal is saying and their own lives. Once the people change, the animal's behavior often follows.

Sometimes, one contact is enough to resolve the situation. In other cases, it takes several. I suggest check-in and follow-up sessions as they seem necessary. Some people like to do quarterly or semi-annual check-ins as a preventive measure.

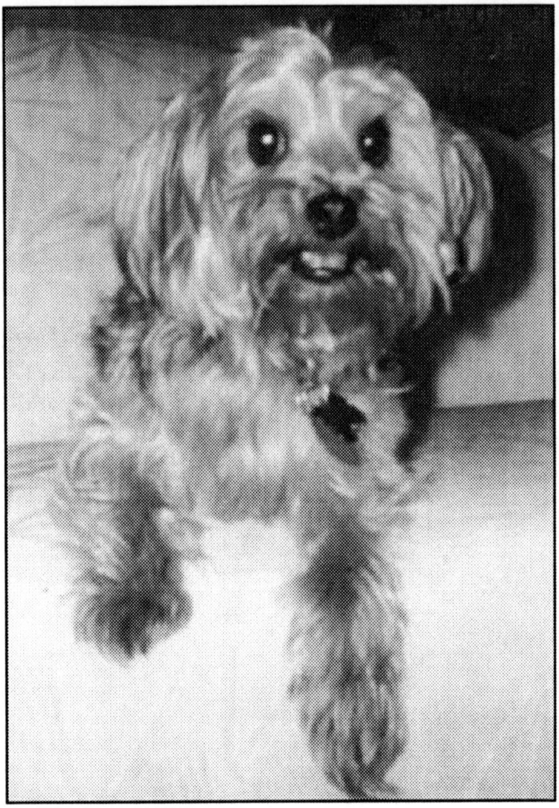

Jake

"I Want a Real Name."

Robin named him Merlin, because he was going to be magical in her life. Only a month before, she had lost her beloved dog Buddy to a misdiagnosed fungal lung infection, and she was not looking for another dog.

Buddy was an eight-year-old terrier mix who had been with Robin through a series of medical crises. She was also studying at a spiritual science school, and he would lie next to her as she meditated or read. When she began counseling clients, he would sit next to her and lean into her leg at critical times during a session. She had come to trust and rely on his input. She was devastated when he died.

A friend told her about this five-year-old yorkie who had been at a shelter for several weeks. His former people, who had doted on him, were elderly and could no longer care for him.

He captured her heart immediately, in the way that dogs do. When she met him at the shelter, he danced at her feet, begging to be picked up. He was acting like a puppy, running laps around the room in a frenzy of canine excitement.

She wasn't sure it was a good idea. If she were to get another dog, she wanted one who would assist her in her counseling practice, as Buddy had done. She didn't know if she could avoid comparing him to Buddy.

The yorkie had touched her, though, and, despite her reservations, she scooped him up, negotiated his "release" from the shelter, and called him Merlin.

When she got him home, his behavior changed almost immediately. He began snapping and growling—at her and at anyone who came around. He was afraid to go out in the backyard, which backed up to a busy thoroughfare and was long, narrow, and noisy. He'd go on a long walk and then come inside and poop in the house. Robin didn't know what to do. She wanted to keep him, but was reluctant to get attached because of his aggressive behavior.

With this as background, I asked Robin to relax and tune in to Merlin.

He had a lot to say. I could feel his anxiety and desperation.

He said he was frightened of the constant noise in the backyard. Robin confirmed that the noise in the yard was continual and loud.

Then I sensed confusion and deep sadness from him. I sent the feeling of sadness to him and thought to him, "Why are you so sad?" Then the thought came back: He didn't understand what happened to him. He loved his first people, and they just left him without saying goodbye or telling him why they were leaving him. He was relieved to have a new home, but he was afraid he'd be dumped again.

I was not surprised by this information and had even anticipated it. I have found that the reason for a sudden behavior change often stems from this kind of sudden abandonment. It frequently comes from people underestimating an animal's awareness and intelligence, believing that if you don't talk about it, the animal won't know what is going to happen.

Intuitively, I knew there was more to this story. I had a felt sense of incompleteness. I asked him what else was upsetting him.

Then he said he wanted a real name, like Sam or Max.

Early on, I would have thought I'd made that up. However, I've learned to trust that the most surprising or unexpected comment from an animal is very real. They have little need to elaborate. Usually, they tell us what is important with a certain kind of intensity, and then we either get it or we don't. They'll tell us again if we ask, but if we miss it or ignore it, they rarely belabor it with us.

When I told Robin that Merlin wanted a different name—like Sam or Max—she said they were the names of two dogs in the neighborhood, whom Merlin had met.

Robin said she had suspected that he felt abandoned. "I think I want to keep him," she said, "but I don't know if I can. If he's going to growl and bite, I can't have him here with my clients and the groups I run."

I asked her what she could tell him that was true. It was crucial to tell him only what was real. If she wasn't sure about something, wait until she was.

She said to tell him he would be with her forever if the growling and snapping stopped. She agreed to tell him what his routine would be each day so he wouldn't be surprised. And she would change his name to something he liked better.

We tried out several "real" names and he let us know he liked Jake the best. Robin reported that he wagged his tail intensely when she got to that name. At the same time, I felt joy and relief from him.

I reminded Robin that she could talk to him any time she wanted to—focus on him and call him silently three times. She said she would do that. She called later that week with a status report.

She said she told him there was room in her heart for both him and her memory of Buddy, and there was room in his heart for both her and his former people. I thought this was a stroke of genius.

Jake, on his part, agreed to think seriously about the whole situation of his new life. It took about six months for him to come to terms with it.

Robin says that Jake has developed an uncanny ability to zero in when a client is distressed. "He nudges me, like Buddy did; but he will also whine and bark in what I interpret as understanding and compassion. Sometimes he'll run around and jump on things like he's two years old, which always lightens things up. His timing is impeccable, and he's rarely wrong."

Jake tells me he is happy with his job. On rare occasions, he still shows hints of the old aggression. He'll growl deep in his throat, or he'll poop in the house. It's usually when Robin has been distracted, or has ignored him too much. It's his way of letting her know he's displeased.

Jake's Message:

We always know when something is going to happen to us. We feel what you feel, and see what you picture in your head. You won't fool us by pretending nothing is going on.

Yofey and Tiger

"Tiger's Being Mean to Me."
"I'm That Important?"

Yofey was ten years old, somewhat neurotic, and devoted to Judy and her son Ben. She had a few quirky habits, but who doesn't?

Yofey was a twenty-pound black and white mix of dozens of breeds. From the time Judy and Ben rescued her from the local Humane Society, she was "Velcro dog," staying so close that Judy had several close calls tripping over her. When Judy and Ben prepared to leave the house, she would give them the most practiced depressed dog look, and then go hide under a bed or in a closet.

When people came to visit, she would quiver so hard with excitement, she nearly levitated. She would lie on the back of the sofa like a cat. Fortunately, Judy is a therapist who works from home, and knew how to manage some of Yofey's more obsessive behaviors.

Life went along in a comfortable routine. Yofey would greet clients coming for sessions, and then sadly slink away when she realized they were not there to amuse her.

She would visit neighbors when she went on walks, play intensely with Ben, and nuzzle up in comfortable companionship with Gingi, the fourteen-year-old orange tabby who had taught her the ways of the house when she first came to live there.

When Gingi died, Yofey grieved deeply for him, as did Judy and Ben. Yofey had never known life as the only animal in the house, and was struggling to figure out this new role. Then, after a few months, life shifted—radically.

Tiger came into the household, and Yofey plunged into a maelstrom. Tiger was energetic, free-spirited, macho—and ten weeks old. He was full of himself, as kittens often are.

A few months after life with Tiger began, Yofey suddenly—and uncharacteristically—started peeing in Judy's bed. Her favorite spot was the middle of a down comforter.

When Judy called me, she said she had tried kindness, discipline, yelling—and still Yofey peed on her bed.

We scheduled a phone appointment. When I tuned in to Yofey, she was clear about the problem. She sent a mental picture of being on a leash, going for a walk with Judy, and Tiger jumping out of a bush. I got the thought that Tiger was being mean to her and no one stopped him. I also got the thought that Tiger taunted her about being on a leash when he could run free.

Judy said she was oblivious to this feline ambush.

I told Yofey other ways to let Judy and Ben know about the torment (whining, staring, knocking over the water bowl, etc.). Then I suggested we turn our attention to Tiger. It took several conversations to get Tiger to understand about living in harmony.

Connecting with him was a little more difficult than with Yofey. He was still full of kitten frenzy, and had the attention span of a gnat.

In the first session, after greeting him and admiring his hunting prowess (he let me know he was very good at it), I asked him how he liked Yofey. He clearly thought she was a fun playmate. I told him that ambushing Yofey was unacceptable, and sent him the thoughts and mental pictures of what was okay: accompanying them on the walk, telling Yofey what he had seen while he was out doing his cat thing, running ahead of them and scouting the trail.

Tiger tried his best, but after two days of being the perfect cat, Judy said he came tearing down the steps at 5:00 a.m. and burst through the patio screen. I sent out a mental instruction for him to come home, and told Judy to call when he did. He was gone for twelve hours.

We set up another phone appointment shortly after Tiger's return. I told him Judy and Ben missed him when he was gone, but I got a sense that he really didn't understand what that meant. I decided to let that go in the moment, trusting that we would find a way to have it make sense to him.

The next thing we tackled was the specific behavior Judy desired. I sent him the thought and mental picture of how the house felt when he was calm and playful, and how it felt when he was raucous and taunting—emphasizing that playful was best.

Raucous and taunting, however, seemed to be Tiger's preferred way of being.

I suggested to Judy that she do an appreciation exercise with him: to get into a quiet, loving place inside herself, tune in to him, and tell him what made him special and his qualities that she loved the most.

She reported that Tiger seemed to calm down some—a very small some, but some.

Finally, we worked on his place in the household—sending him the sensation of how it felt when he was there, and how it felt when he was gone. This time, he got the whole picture. His reaction was a feline version of incredulous. I could feel his surprise. It was like an "ah ha" in a person. He had no idea how important he was to the family. This is a highly abstract concept for a cat, and it speaks to Tiger's intelligence and evolution that he understood it.

Yofey stopped peeing on the comforter after the first conversation, and listened in—whether invited to or not—on the subsequent talks. Her tension was palpable; she was ready to tell on Tiger if he made even one transgression.

They developed an easy, comfortable relationship over the next seven years. As the infirmities of age took their toll on Yofey, Tiger became very solicitous. He would stand watch until a seizure ended. He would nudge her around a table when her dimmed vision led her to a near collision with rearranged furniture. He would take an afternoon nap curled up on a sofa a few yards from her.

At the venerable age of eighteen, Yofey died peacefully at home. Judy said she watched Tiger say goodbye: He sniffed her, and then, for about fifteen minutes, he lay next to her in quiet reflection.

Yofey's Message:

Don't you know that we talk to each other all the time? We think to each other in our heads. You just don't know what we say.

Tiger's Message:

Tell us how we fit into the family, and what you expect of us so we know what you want.

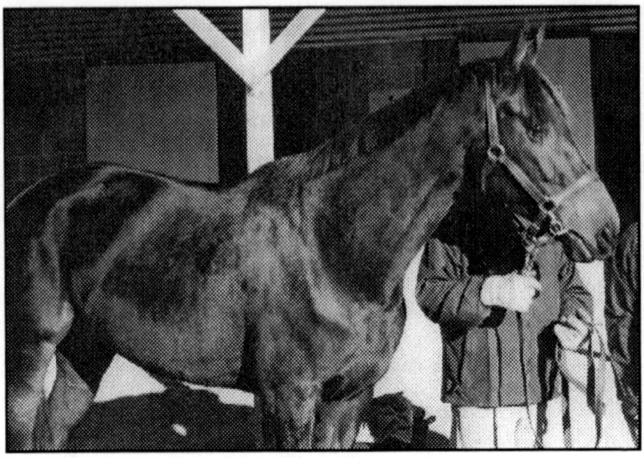

Joanie

"I Want to Be a Mom."

Joanie is a beautiful, well-bred racehorse who was greatly underestimated by her breeder. He thought she wasn't fast enough to win races, so he set her up as a "rabbit"—or pacesetter—for other horses training at his stable. "Rabbits" in the horseracing industry are horses who start fast and then lose their momentum. They set the pace for the other horses being trained, and, as their momentum slows, the other horses pass them. The term comes from the dog racing industry, where a wood-carved rabbit sets the pace for the competing greyhounds.

Joanie's owner decided to sell her because she wasn't up to the level he preferred his horses to be.

An investment partnership was looking for something to purchase, and bought Joanie because the price was right. Frances, a friend of one of the partners, bred and trained horses herself. Frances recognized the spark of talent Joanie occasionally showed in spite of her pattern of fast starts and then losing momentum. She took on Joanie as a project.

At Joanie's first race with her new owners, she was so stressed that she reared up and fell over backwards while being saddled. Frances was savvy in the ways of conveying information to her horses, but couldn't get information back from them. She needed a dialogue with Joanie. She planned a covert mission with me to the track; she didn't want to deal with the stable hands and trainers who were disinclined toward animal communication.

When I tuned in to Joanie, I felt both physical and emotional pain.

Doing a psychic scan of her body, in which I aligned my body with hers and then felt what she was feeling, I found that her back hurt a few inches from where her saddle ended. I felt tightness and pain in my middle and lower back. She showed me a mental picture of something hitting her there. It looked like a tree limb. A veterinarian (who is also an acupuncturist) later confirmed the injury and treated her for it. Her right front ankle, which she had injured as a yearling, was fine, she said, as long as she could warm up.

Emotionally, however, Joanie was a mess.

She was confused. In her training, she always started out first, and then the other horses passed her. She would be praised and rewarded for this. Then she overheard humans saying she was a "rabbit." She thought she was a horse and was supposed to let the other horses go past her.

The stress of racing—all the excitement and noise, anxiety, and frenzy of the track—also upset her. She would get anxious, similar to performance anxiety in people. She would lose her focus. People would yell at her and she didn't know what she was supposed to do. The other horses would talk to each other about who was going to win. She wasn't sure what winning meant.

Frances, who owned a brood mare farm, had an intuitive thought that Joanie might want to be a mom. She had spent years with mares and their foals, and sensed in Joanie the same desire she felt in her mares when they were ready to be moms. As we visualized Joanie with a foal, and asked her if she would like that, she flooded us with feelings of joy and contentment.

Joanie's genetic line was impeccable. In order to be mated with a stallion of equal caliber, she had to win a race. The problem: Her idea of winning had nothing to do with racing. She thought winning meant squeezing the other horses out of the feeding bin.

I showed Joanie a mental picture of crossing the finish line in front of all the other horses. I also told her how sorry we were that her original people were so blind to her talent. I told her she was fast and smart and could win if she wanted to.

Frances was open to doing whatever was necessary to help this horse. I suggested she begin an intensive coaching process. Frances practiced sending her concrete, uncluttered mental images, such as crossing the finish line ahead of the other horses. I checked in with Joanie to see if she understood Frances. It did not take very long for Joanie to understand.

Then, before each race, Frances would walk with Joanie from the stable to the paddock, visualizing to her what winning looked like. She also sent Joanie the picture and feeling of being relaxed, and of keeping all four feet on the ground.

In her first race with her new owners, Joanie came in second. She repeated that performance in two more races. And then, she won. Her owners were ecstatic. Joanie took her place in the winner's circle with a sense of pride and grace that told onlookers she knew she belonged there.

It was the only race she ever won. The group raced her the rest of the season, but she never finished above eighth place after that.

At the end of that season, she was retired from racing, and spent many years in the bluegrass of Kentucky. Her first foal was sired by Belmont Stakes winner Victory Gallop (who also, by the way, placed in the Preakness and Kentucky Derby).

Joanie's Message:

We remember what you say, even the cruel things that you may not mean. If you want us to be our best, tell us what our best is. We will believe you.

Gambit

"What Happened to My Family?"

"How can a bird be expected to understand divorce?" That was the question Linda asked when she called, desperate to understand the radical change in the behavior of her beloved parakeet. She and her husband, John, had separated after a particularly nasty incident, and both were trying to regroup and re-establish their lives.

They agreed that Gambit, whom they both loved dearly, would stay with John while Linda settled into her new place and prepared for Gambit to live with her. Gambit spent six weeks in this transition. During this time, he picked at his feathers rather compulsively, a sure sign of anxiety in a bird. He had not picked himself raw, though.

Then he went to live with Linda in her new home. He would either nip and bite her, or ignore her completely. Gambit had always been a very social bird, and had never been aggressive before. At the same time that he was becoming increasingly aggressive toward her, he would greet people coming to the house, sit on their shoulder, and be generally friendly and loving to them.

Linda was absorbed in adjusting to her single life, and had no idea how to fix her relationship with Gambit.

As I was tuning in to him, I could feel his angst and his eagerness to talk. Gambit quickly laid out the whole scenario for me. He had always been with Mom and Dad and then he was in a new space with just Dad. Then he moved to another new place and just Mom was there. No one told him what was happening. He loved his family and they weren't together. He was confused: He thought he had done something to make them split up. (Without input from us, our animals will reach their own conclusions about the situations they experience.)

He didn't know what they wanted him to do, and he wanted to know why this was happening to him.

I explained to Gambit that one of the ways humans are different from birds is that sometimes humans get so angry or upset with each other they can't live in the same house anymore. When this happens it has nothing to do with the bird in the house.

"Mom and Dad love you," I told him, "and they want you to live where you'll be happy."

I suggested to Linda that she tell Gambit he was a keeper. I sensed that he needed to hear it from her, and not through me, so I coached her in how to connect with him—and then stayed connected to make sure Gambit got it. Linda told me she sent these thoughts to him: This was his forever home; she wanted him to be with her if he wanted that, too. She apologized to him for disrupting his life and home with no warning. She also agreed to tell Gambit when things were happening and how he would be affected.

"It was like magic," Linda reported. Within minutes, Gambit flew to her and settled on her shoulder, chirping and rubbing his head on her face. Sometimes, when an animal knows he's been heard, his behavior quickly returns to its pre-tantrum state.

Within a couple of days, Gambit resumed his imperious presence, perching on her head and squawking at her when she was on the phone.

As Linda recognized how sensitive Gambit was, she began telling him about the things that were happening in her life, both good and bad. Over time, Gambit fell into a routine of greeting her with a squawk and a song whenever she came home, stopping only when she greeted him in return and told him about her day. When she forgot to tell him, he would impatiently chitter at her until she did. They both settled in comfortably to this new arrangement, and Linda has learned to detect his signs of distress before Gambit has to go to extremes to get her attention.

Gambit's Message

We know when you're sad or scared or angry, and sometimes we think it's our fault. We don't know we didn't cause your problems unless you tell us.

Mazik and Coco

"Too Many Changes."
"See Me Too."

Mazik and Coco are both chihuahuas. They are not littermate brothers, but you would never know it by the way they get along. They are devoted to each other. By mutual agreement, Mazik is the leader and Coco his loyal shadow. They love car rides and long walks, and are happiest when they spend all day with their person, Nancy.

Mazik was always proud of his ability to safeguard the house. In spite of being only eight pounds, he made sure everyone knew he was a huge force to contend with.

Shortly after they moved into a brand new townhouse, Mazik changed.

Nancy called me in desperation. That's often the way people get to an animal communicator. Things are going along just fine, and then there is a significant change in the behavior of their animals, and they have no clue how to fix it. They often seek veterinary care, and just as often the issue is not medical.

"I don't know what's wrong with him," Nancy said. "He's just not happy. His tail has been down for weeks and he's peeing all over my new house," she lamented.

Coco was clearly worried, and showed it by increasing his entertainment behavior—running in circles and tossing balls in the air and chasing them. Nancy thought this was beyond his usual attention-getting ploys.

She didn't know what to do to help her boys, and felt like "an awful mom."

Nancy requested a home visit instead of a phone consultation so she could see their reactions to me. I explained that the results from using a photo and the phone are just as accurate and, because it takes less time, less expensive.

"I'd rather you come here anyway," she said. "I've got to tell you, I'm a skeptic about this stuff, but I'm desperate and I'll try anything. I want to see their reactions."

She warned me that Mazik was not a friendly dog.

Within ten minutes of my coming into the house, he raised and wagged his tail. He brought me his ball. He brought me a Frisbee. He brought me his favorite stuffed toy. He would only leave my side or my lap to chase a ball. Coco spent that time jumping on and off the sofa like a clown.

Mazik is usually a very serious boy. I could sense that in his attitude and thoughts. I also got a feeling of instability and frenzy from him, related to loud noises. He showed me a mental picture of a grassy yard, and something in the ceiling in the kitchen, which turned out to be a fire alarm.

He said there were too many changes. A lot of things happened really fast. The yard hurt him. Nancy was suddenly gone a lot and it wasn't the same with the dog sitter. He was eating a treat in the kitchen when the thing in the ceiling made a really loud and shrieking noise that went on for a long time. It hurt his ears and made his skin burn. He was scared and upset that another big noise would happen. He couldn't do his job of protecting the house.

He was noticeably relieved after he'd had his say. The worry lines around his eyes smoothed out, and his body relaxed.

Nancy was amazed at his physical reaction, even before she knew what he had said.

She confirmed that he had been zapped by the newly installed electric fence, and he was standing in the kitchen when the smoke alarm went off. She also told me she had been gone more than usual, between a heavy work schedule, getting the new house in order, and a long-planned vacation.

I told Nancy she needed to tell him what was going to happen, so he wouldn't be surprised. She asked me to tell him that her traveling would slow down, and that the cause of the noise was fixed.

Coco, meanwhile, was confused by Mazik's behavior and showed his anxiety by increasing his cuteness. He didn't know what to do when Mazik was so upset. He thought he could distract both Mazik and Nancy so things could settle down at home.

We reassured him that life was settling down and that being an entertainer was a very important job in the family. Trying to explain "comic relief" to a dog is tricky. I sent him a mental picture of Nancy laughing and relaxing at his antics. Coco got the idea, and he, too, visibly relaxed.

So, Mazik got to talk about his experience, Coco got validation for his role in the house, and Nancy felt "incredible relief" that she had an ally in raising and caring for her boys.

"I won't consider anything major without talking to them," she said. "You are our intermediary. You can tell them what's going on when I don't think I'm reaching them, and you can interpret their behavior and direct me to a vet, to a trainer, or back to myself and my own instincts when I'm too close to the situation and lose contact with my own intuition."

Mazik is still somewhat wary. One of the things he wants to do in this life is to learn how to adapt to changing circumstances. He knows he picked the right human.

Mazik's Message:

Wouldn't you pee, too, if a big noise scared you and hurt your ears and skin and you couldn't do your job?

Coco's Message:

We are who we are, and we do what feels right to us. Let us know that what we do is important.

Kiwi

"I'm Part of This Family Too, You Know."

The older Kiwi got, the cleverer she became at running her family a merry chase. Before leaving the house, they would run after her for ten to fifteen minutes to catch her and put her in the laundry room. Otherwise, she would pee anywhere in the house that she chose.

The daily chases were wearing thin, not to mention how late Maria, Ken and their three teenagers were for their various appointments and jobs. The frustration drove them to wonder if Kiwi were just playing with them, or if she had a message.

Kiwi was just waiting for the opportunity to tell her side of the story. She ragged on them. She complained about all the people in the house coming and going and ignoring her. She wanted a lot more attention than she was getting. She said, "I'm part of this family and they don't talk to me."

Kiwi was very bonded to Maria, whom she considered her primary person. The kids were now teenagers, and Maria was becoming increasingly involved in a part-time job and other activities that took her away from home. She gave the kids instructions, and left Kiwi more and more in their care.

Maria was committed to keeping lines of communication open with her husband and children. Their house was the gathering place for the kids' friends, and she became the unofficial advisor and mentor to a slew of teenagers. When she heard Kiwi's comments, she was amazed to learn that Kiwi was so aware. She was also appalled at her insensitivity in diminishing her canine companion.

Kiwi understood she couldn't go with them whenever they went out, even though she wanted to. I suggested to Maria that she tell Kiwi exactly when someone would be home. Animals don't tell time by the clock; however, each hour has a feel to it, and if you say you'll be back at 3:00, you will be holding a mental picture of 3:00, and the feel of 3:00. Our animals tune in to that "feel," and when that feel happens, they know we'll be home. It gives them a sense of something other than the here and now, and has worked well with mild separation issues.

Kiwi also wanted a choice about where she stayed when she was left alone.

Maria was fascinated by this animal's acumen. I suggested she tell Kiwi how smart she is, and what makes her so special. When we talk to our animals this way, and couple it with feelings of love and appreciation for them, they get the whole message.

Kiwi responded immediately. She stopped playing "keep away," chose her place to stay, and stopped peeing in the house.

Kiwi started coming into her own after Maria's recognition. She would maneuver between the kids when they argued. She would expectantly anticipate being invited in when company was over.

She knew the boundaries of the property, and would pee or poop on the property line or on their side—never on the neighbor's. She wasn't leashed when she joined the family outside because she never ventured away (even when tempted by a rabbit). She would stick her nose by the door and then look back quizzically an hour before anyone left the house. She even started leading the way to the laundry room when they went out.

Maria pointed these antics out to the kids, and they began telling Kiwi how smart and helpful she was. The family welcomed her to their gatherings, and she settled into a comfortable routine.

For the rest of her life, Kiwi fully explored her role as family member. Maria is an energy healer; she would invite Kiwi in to help amplify the energy during sessions. Kiwi also became increasingly sensitive to anyone in the family being in distress, and would sit on their feet until some inner sense told her it was okay to move. She would drape herself around Maria's head at night, or claim part of the bed as hers (leading to what Maria described as "two queens fighting over rightful places in the bed").

I asked Kiwi what her life purpose was. The word "relationships" popped into my head. Through her behavior and attitude, she was here to teach her human family how they were relating to each other, and how to effectively relate to her. She hopes they continue learning this for a long time to come.

Kiwi's Message:

We want to be included and know what's going on in our family. Don't think we are less than you just because we're different from you.

Landhi

"I Want to Stay Aware."

Landhi was diagnosed with feline AIDS when he was nine years old. In his earlier years, he was known as the wild man of the neighborhood. He had two passions: hunting and peacemaking. Every year and a half or so, he would go off for days at a time and come back "with a smile on his face" according to Kristine, his human companion. At first, Kristine would search the neighborhood for several days and then Landhi would just show up. Then, she became accustomed to his routine and would wait a few days before starting her search. As if he were reading her mind (which, we know, our animals do), he would come home just as she started looking for him.

He would gather all the cats in the neighborhood into his territory, and then hold court. Landhi would receive and welcome each new arrival. Then they would bask in the sun together, feet tucked in, in the way contented cats do. Occasionally, another cat would challenge him. He told me he didn't like to fight, but wouldn't back down if one were inevitable.

When he was younger, he had recurrent urinary tract infections, which were ultimately managed by surgery and a special diet. Then he developed a series of symptoms involving his skin, and the fur on his rear end fell out. The vets ran a battery of tests, and they all came back fine—except the one for feline AIDS. They gave him six months to live.

Kristine, a strong proponent of alternative healing for herself, sought out alternatives for Landhi. First she consulted with a homeopathic vet, who was surprisingly optimistic. The vet viewed this kind of illness not as a death sentence, but as a wake-up call to the need to make some changes. Kristine was relieved to find there were options, and this vet became Landhi's primary doctor.

Several of the options the vet suggested were essentially illegal in the state where Landhi lived. This was at a time when alternative healing for animals—acupuncture, energy healing, massage—was a highly political issue, and practitioners were working in secret.

Kristine tapped into this underground network, and would hustle him into his carrier for clandestine visits to the undisclosed offices of an acupuncturist and an energy healer.

Landhi had an up and down experience after that. Periods of dire illness, in which he was lethargic, not eating, and feverish, were followed by periods of seemingly robust health. Kristine always worried that he would not pull out of an episode, but he always rallied.

As a devotee of yoga and meditation, Kristine had created her home to be a sanctuary from the outside world. An altar figured prominently in the décor. When Landhi was sick, he would somehow pull himself up onto the altar and sleep. When he was feeling good, he would hold court outside, sleep in the garden, and run to greet her when she got out of her car.

When Landhi was diagnosed, Kristine prayed that she would know what to do to support him. She had never lived with a cat with a chronic illness, and she called me to help her trust her instincts with him.

I had many conversations with Landhi over the years of his illness. Sometimes Kristine would just want to check in with him. Other times she suspected he was about to go into a flare-up and wanted me to assess his health. On occasion she had specific concerns she wanted to discuss with him. She described it as "good couple's counseling."

Every time I connected with him, I felt him growing wiser. He was a role model in adapting to a chronic illness with graceful acceptance. Near the end, when I could feel his life force diminishing, he said he wanted to stay aware through every minute of his dying process. It was his time, he said, and Kristine should not worry when he stopped eating.

She began a life review with him. She reminded him of the highlights of their life together, and told him very specifically what made him so special and remarkable. The more she did this, the more sure she became in knowing just what to do for him, and when to do it.

Landhi lived for nine years after his diagnosis. He died peacefully in Kristine's arms. She described it as one of the most profound, loving moments of her life.

Landhi's Message:

We have our own ideas about the quality of our lives and ways of dying. Ask us what we want. We will tell you. We're not afraid to die, and we know we can come back in another body any time we want to.

Sasha

"Please Understand Me."

It's hard to believe a ninety-pound boxer would be afraid of anything. However, Sasha has always been a sensitive girl. She would shake uncontrollably and curl into a tight ball in the back seat of the car whenever an eighteen-wheeler passed them. The vibration made it feel to her like the ground was falling apart.

She is very connected to Nancy and Bob, and a rash pops out on her belly when one of them is upset or when they argue. At the same time, she can be a fierce watchdog. Any animal or person approaching her house knows she's on duty.

Sasha had been a regular at doggie daycare for daytime play. She loved the place. For years she recognized the route, and whined and yipped in anticipation when the car turned onto that street. The staff told Nancy and Bob many tales of Sasha and her "boyfriends."

When Nancy and Bob went on vacation for a week, they naturally decided to board her there.

When they picked her up, though, she was a mess—nervous, afraid, needy, and clinging. She wouldn't leave their side. Always sensitive to noises, she now hid under whatever was available at any unusual sound. They were worried that she had been mistreated, and they called me to see if Sasha would tell us what happened.

Sasha said she was trapped in her cage when a loud noise started, and she couldn't escape. The banging and pounding went on for a long time. The sounds not only frightened her, they also caused stabbing pain all over her skin. As if that weren't enough, she said something got in her cage and she couldn't get away from it.

Once she got home, Sasha was even more upset. She couldn't do her job of protecting the house because noises sent her running for cover. Her belly was a mess of rashes.

Nancy and Bob checked with the daycare staff, who confirmed that there were severe thunderstorms in the area while Sasha was there. They also found that the roof of the kennel is aluminum, which amplifies the sound of battering rain. The staff didn't know if anything had gotten into her cage.

It's taken a while for Sasha to regain her courage. Animals do experience the equivalent of post-traumatic stress, and allowing them to tell their story helps resolve lingering fears. Sasha and I had several conversations. She showed me mental images and her feelings of helplessness and terror at the unrelenting noise, along with her sadness that she was disappointing Nancy and Bob. I told her that she did all she could do, that there was no way to get away from the noise, and that Nancy and Bob were worried because she was so upset. I also told her I was so sorry that had happened to her.

I then coached Nancy and Bob to tell her that she was still the wonderful companion she always was, and to reassure her that she will never again have to stay in that place overnight.

Between these conversations, patience on the part of her people, and a holistic vet who provided acupuncture, herbs, and flower essences, Sasha is handling the noises of fire trucks, ambulances and eighteen-wheelers much more bravely.

Her vet informed Nancy and Bob that extremely sensitive hearing, like Sasha has, is connected to fear. As the fear is managed and released, the ear sensitivity decreases.

Nancy and Bob began telling her what would be happening that day. They told her in advance when they knew there would be unusual noises. They encouraged her to patrol the house as she had done before. They rewarded her for chasing other animals out of the yard.

Over time, Sasha's rashes have improved significantly. She sits relatively calmly on the car seat when they go for a ride. She still loves daycare—as long as she gets to go home that night.

Nancy and Bob have been able to find other caretakers for Sasha when they travel.

They know, however, that the condition of her skin will remain a barometer for their own fears and conflicts.

Sasha's Message:

When we can't get away from loud noises or mean voices, we feel trapped. Find out what scares us or makes us angry, and then you'll know why we do what we do.

Pocca

"You Just Don't Get It—My Shoulder Hurts."

Pocca is a purebred golden retriever, and she had been a stellar student for over a year. She excelled in puppy obedience. She did well in basic agility and, even though she was skittish on the seesaw (she has never liked things that bang), she loved the ease and grace of her body as she weaved, scurried, and balanced on the other equipment.

Agility is a competitive sport in which dogs, at their person's direction, race through a course of physically challenging events: going through tunnels, balancing on boards, crossing ladders, weaving between poles, jumping over obstacles. It requires a close partnership in which the person "reads" and directs the dog, and the dog anticipates the person's direction. It is supposed to be fun.

A couple of classes into intermediate agility training, Pocca simply refused to do anything. When Deena, her person, tried to coax her to participate, she slipped out of her collar. Baffled by Pocca's obstinacy, Deena jokingly told her she was being a "juvenile delinquent."

Then Pocca did some odd things. At home that night, she refused to snuggle with Deena, a routine they had both enjoyed since she was a small pup. She also pointedly avoided Deena, staying on the opposite side of the room and staying outside for hours at a time. She was obviously miffed, but Deena had no idea why. After a week of this behavior, she called me to find out the reasons.

Pocca was incensed. She also didn't want to tell me why. When I asked her about the pain I felt in her shoulder, she showed me herself jumping over an obstacle and landing off balance and hard. Then the thought of "bad name" came into my mind, along with a feeling of embarrassment.

When I relayed that information to Deena, she told me about the "juvenile delinquent" remark. Pocca, who was listening in, got angry. I was getting a clear message that Pocca's hurt was emotional as well as physical. She told me she hurt her shoulder and Deena didn't even know it. She always knew what was going on with Deena, and expected the same in return.

I told Deena that while many dogs—and Pocca is one of them—have wonderful senses of humor, they don't understand exaggeration. A chagrined Deena apologized to Pocca for being so insensitive to both her feelings and her physical pain. We explained to Pocca that she showed no signs of being hurt—she didn't yelp or limp. She just refused to move, and Deena mistakenly thought she was being stubborn.

Pocca, however, was not ready to be consoled. She was reluctant to give up her anger, and, in effect, did the doggie equivalent of rejecting the apology.

It is not in a dog's nature to hold a grudge. The fact that Pocca was still so angry a week after the incident showed both her sensitivity and dogged persistence. The pain in her shoulder was gone, but she wanted to be sure Deena got the message.

Then it took about twenty minutes to counsel Pocca out of her anger. We showed her the feeling of what life would be like if she stayed angry: Her heart would be closed off, she wouldn't feel closeness with Deena or other people, and she would feel sad or angry. We also showed her what life would be like if she forgave Deena: close connection, open heart, happy. She decided that staying angry didn't feel good.

Deena reported that Pocca resumed her "maximum snuggling posture" that night, and they have been fine with each other ever since.

This was a crash course for Deena in being careful about the words and feeling behind her communication. Now she is prudent about what she says to Pocca. She has also honored Pocca's request to stop agility classes.

Pocca's Message:

We see what you are thinking* and we know how your thoughts feel—the bad ones and the good ones. It feels good when you apologize for something you did.

*Author's note: Humans think in pictures as well as words. Our animals tune in first to our feelings, then to the pictures of our thoughts, and finally to the words that we use. Our animals also know if our thoughts feel good or bad to them.

Zoo Bird

"Stop It!"

Zoo Bird had a difficult early life for a cockatiel. He lived in a place where he was nervous and upset all the time. The humans argued a lot, loud music was blaring long after his cage was covered for the night, and a cat chased him whenever he flew around the house.

For reasons he couldn't—or wouldn't—reveal, he wound up in a bird rescue.

Debbie, an animal behaviorist who is gifted in working with abused animals, then adopted him. She added him to her rescued flock of six other parakeets and cockatiels.

He came to his new home a nervous wreck. He had picked himself nearly raw. He was more easily startled than most birds, and he was wary of humans. Yet he had this endearing habit of interrupting his normal singing with a mean rendition of "Dixie" (you know, "Way down south in the land of cotton . . .").

The birds lived in Debbie's office in large, well-appointed cages that simulated their natural habitat.

Under her careful handling, Zoo Bird began to relax and greet her with "Dixie" each morning. He also quickly learned to imitate the sound of Internet dial-up, and would regale her with his own unique concert.

The office was the site of many meetings, and Zoo Bird became the official welcomer. One of these meetings got heated as the people became increasingly frustrated in planning a major fundraising project. Voices got loud. The atmosphere got prickly and intense.

Then Zoo Bird chimed in with a cacophony of noise that sounded like a frenzied, off-key version of "Dixie" with a few phone line beeps thrown in. He also picked at his feathers between squawks. He got louder than the people in the meeting. He got so loud, in fact, that the people stopped arguing.

After a break, the meeting resumed with a milder exchange, and Zoo Bird returned to his melodious comments.

Debbie was curious. She wanted more information than she could get from observing him.

In my first conversation with him, I was impressed by how sensitive Zoo Bird was. He was more aware and more sensitive than is typical for his species and breed. He told me about the place he had lived before, which left him feeling jangled and on edge. When the humans were here, it sounded like where he used to live, and he wanted the noise to stop. He was yelling at them in the only way he knew.

I suggested to Debbie that we find a way to use his unique talent.

I told him he could have a very important job. The people who gathered there were all helping animals in some way. If he would squawk before the situation got too intense for him, the people would pay attention and settle down. This would let them find more ways to help animals.

Zoo Bird has taken to his job quite well. Debbie tells him just before a meeting is going to happen, and reminds him that she will pay attention when he first starts to squawk. She tells the people attending the meeting that Zoo Bird will keep them on track.

He is now a permanent fixture at all meetings. The people are well-trained. When he starts squawking, they stop arguing. Sometimes they calm down and lower their voices; sometimes they take a break and resume the meeting in a more thoughtful attitude. Zoo Bird has become an excellent facilitator of harmonious group dynamics. The people are pleased with the clarity of his interventions.

Zoo Bird's Message:

We like it when you listen to us before we have to do something you think is really bad. We try lots of ways to tell you to stop when it doesn't feel right.

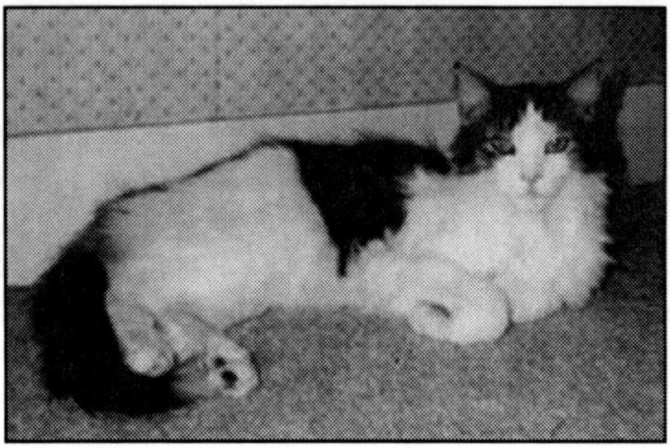

Sunny

"I Want to Be Free."

Sunny is a magnificent silver-black-white domestic longhair cat with an attitude.

When he was about five years old, he suddenly began peeing outside of his litter box. On one occasion, his person, Lee, caught him peeing on the living room carpet; another time, she saw him walking away from a puddle in front of the kitchen door. Usually, though, while olfactory evidence indicated that one of the six cats in the house had peed outside of the litter box, the actual culprit was not obvious. Lee didn't want to accuse just anyone.

I decided to call all the cats in together. Lee and I visualized each one and called his or her name silently until I felt the connection. I told them that Lee wanted to talk to them about the peeing. Then I projected to them a mental image of Lee discovering the puddles and reacting to the smells in the house. Everyone but Sunny said, "Not me." His silence, though, was deafening.

Holding a curious, non-judgmental attitude, I asked Sunny what he wanted Lee to know. What I got from him were feelings of anger and frustration, with a mental picture of him committing the act.

Animals are not capable of lying. They don't say yes when they mean no. In person, they may have characteristic ways of pretending not to be guilty, but their lack of eye contact, or the position of their ears, gives them away. In a phone consultation, however, when there is only a mental connection, they relay attitudes and thoughts. If they don't want to talk about something, they may tune me out or refuse to comment. At those times, I will assume an answer, and will get back from them either an angry denial or a feeling of their relaxing into the conversation.

I've never had animals say they didn't do something if they really did.

When a cat pees outside of the litter box, it is usually a statement of being angry or scared. Most of the time, they are sneaky about it. If a cat allows himself to be seen peeing inappropriately, it often indicates a physical problem. But Sunny was physically fine, according to his vet.

Resolution with Sunny took some time.

Lee and her husband Larry had recently screened in part of their deck, and the cats loved it out there. Sunny's excitement about being outside was obvious. He said he felt free outside, even though he had been content as an indoor cat for several years. On direct questioning, he acknowledged that he was marking his territory by peeing around the house, as if he were an outdoor cat. He showed me a picture of himself out in the yard. When I asked if he wanted to go all the way outside, beyond the screened deck, I felt a flood of relief and excitement.

After a month of weekly conversations with Sunny, Lee, and Larry, it was clear that Sunny would not stop marking around the house. The choices came down to sending him to a farm, where he could be a barn cat, or letting him be an indoor-outdoor cat with them. Sunny's reaction to the barn idea was a strong and immediate "NO!"

Lee and Larry knew the dangers of letting him be an indoor-outdoor guy, and they wrestled with how to be responsible cat owners and still have the household be livable. With much trepidation, Lee and Larry agreed to let him go all the way outside, but they wanted him to do two things: to stop peeing in the house and to stay very close to home when he was outside. I felt Sunny's agreement. But the promise was short-lived.

Animals respond to things in the moment. What is true for them today may change tomorrow or next week. If they "break their promise" it is not willful defiance; it is their nature to be true to themselves in the moment.

Sunny started roaming farther from the house, and Lee and Larry were getting nervous. Their house was only a block from a major road, and their street had become a shortcut. Sunny had never been around traffic, and they didn't know how quickly he could become streetwise. There were also feral cats and foxes in the wooded area behind their house. Could he hold his own with them? Outrun them? Find a safe hiding place?

We turned to the newest member of the feline family—a streetwise stray who had adopted them about a year before. Lee fed him regularly on the deck, and created a shelter out of a large cat carrier and blankets. When he became ill, they scooped him up and took him to the vet, who ordered two weeks of medication and recuperation. At that point, they took him in and named him Buddy.

Buddy was a big guy—sixteen pounds of grey tabby, a "Don't mess with me" attitude, a mangled right ear—and all heart. He easily adapted to domestication, allowing himself to be picked up, sitting in their laps, nuzzling their hands to scratch his face, and joining them (and several of the other cats) on their bed.

Sunny and Buddy had forged an uneasy alliance. Sunny would alternate between growling and hissing at Buddy, and then grooming his head. Buddy, in turn, would scope out the scene (apparently checking Sunny's mood), and then gently approach him.

I explained to Buddy that Sunny wanted to go all the way outside, and asked him to teach Sunny the ways of the outside world. Sunny, who thought that he needed no advice, became a reluctant student after an owl nearly snatched him for dinner. (Lee called to find out what was wrong with Sunny when he streaked into the house with a wild and terrified look in his eyes. Sunny showed me a picture of dashing under the deck as an owl swooped down on him).

Things went along smoothly for a few months. Sunny and Buddy would occasionally prowl and hunt together, bringing their catches home to share with Lee and Larry.

Then Sunny started peeing in the house again.

Frustrated beyond description, Lee would call and ask me to tell Sunny that he was banished from the house for days at a time because of the peeing. It was happening so often that Sunny would get an attitude as soon as I told him Lee wanted to talk to him. I could feel his anger; he would effectively tune me out and refuse to respond. I knew I was losing my objectivity because I could feel my attitude matching his. If I couldn't stay neutral, I couldn't hear both sides. I knew I was becoming part of the problem.

I called my friend, Deena Spear, who does an unusual energy healing process in which she "hears" the energy field of a person or animal, and then mentally "tunes" the field until it feels harmonious and flowing. I asked her to work with Sunny and me. It was difficult for her to get the harmony, and she sensed that heartbreak was interfering. With another healer on the case, I was able to disentangle myself from Lee's frustration, and could clearly feel Sunny's heartache.

Sunny was guarded with us. "Distrust" popped into my head. I got the thought-impression that the humans in his life didn't understand him, and he didn't have much faith in us. Deena and I agreed that Sunny needed to talk to one of his own kind. We asked a cat named Sylvie, who works energetically with Deena, to talk to Sunny. (I often ask one of my cats to do that when I'm having trouble connecting, but I was too involved in the ongoing drama to remember to do it.)

I felt Sunny begin to relax within a few moments of connecting with Sylvie. Even though I was tuned in, Sunny and Sylvie kept their conversation private. I could feel the exchange, but could not get any detail. My heart connection with Sunny was still open, and after about a minute, I felt him relax with me also.

He was angry and scared that he couldn't connect with Lee the way he always had. He said she was pulling into herself and away from him. He didn't know why. When he was in the house with her and he couldn't feel the connection, he was, in effect, saying "P--- on this!" He was also angry with me for always telling him what Lee wanted, and not hearing him. What a humbling experience that was!

Lee also felt humbled when she heard Sunny's message. She acknowledged that she had pulled back because she thought that was what he wanted when he insisted on going outside. Then, when she was doing what she thought he wanted, he turned on her, and she, too, was angry and upset.

Things improved radically after that. I relayed to Sunny Lee's apology for misunderstanding him, and asked Sunny to let her know when something was wrong by either sitting on her lap and staring at her, or by knocking over a water bowl. Sunny agreed to that.

Lee told me she would call to find out what was bothering him when Sunny did those things. And she did just that. There have been several occasions when Lee has called to ask how quickly she could get an appointment because Sunny has just stared at her, or jumped to the top of the cat condo and won't interact with her.

Sunny's concerns and worries center around changes in the neighborhood, one of his people being ill, or conflicts in the house. He usually settles down once he knows he's been heard.

Lee schedules regular check-ins with all the animals in the house to learn what they have to say. I also suggested she periodically tell each of them what it is about him or her that specifically pleases her. This is easily done by filling yourself with love and appreciation for that animal, and sending that feeling, along with the qualities you love about them, from your head and heart to theirs.

Lee says that things are calmer and smoother around the house now. The peeing has stopped; Sunny and Buddy are spending more time inside than out; Buddy has taken a new kitten under his paw; and Lee is happy that her abilities to connect directly with her animals are increasing.

Sunny's Message:

Once we decide to live with you, we connect to you all the time. When you shut us out, it scares us. It makes us feel all alone. This is not good.

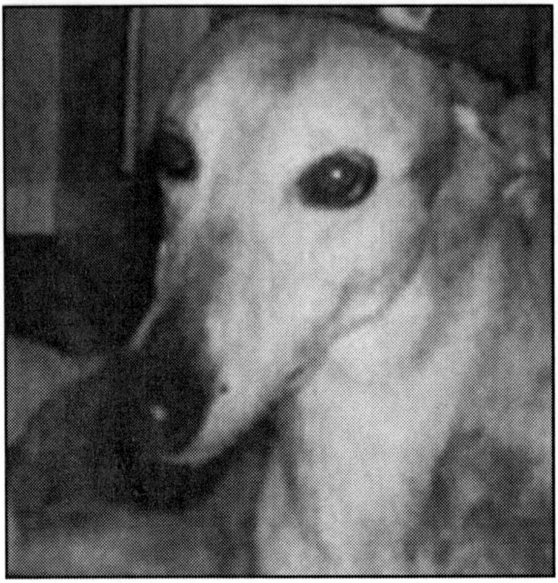

Mae

"Where Am I?"

Mae met her new person, Sendy, under less than ideal circumstances. She was two years old and had just been retired from greyhound racing. She had spent her entire life in a crate, in intensive training, or in competitive racing. She was transported from Florida to Maryland in a crate with another retired greyhound and delivered to a temporary foster home.

Sendy—a single mom of two teenaged boys—had contacted Greyhound Pets of America (the greyhound rescue group) a couple of months previously. She told them she wanted a younger adult dog who got along with cats. Mae fit her requirements, and went to her new home just days after arriving at her foster home.

Mae didn't know a lot of things. She had to learn how to climb steps. She had never seen a window or sliding glass doors. She didn't know how to play, and she was wary of people—especially tall, thin men. She was literally afraid of her own shadow when she went for a walk.

Certain noises and movements frightened her, and she would become instantly alert, ready to pounce. Sendy was afraid that her fear-aggression would go beyond snarling and growling. She called me to find out how to help Mae settle into her new home and new life.

She was particularly worried about two things. Mae would growl and snarl when Sendy's brother came into the house wearing a leather jacket and leather cap. She would exhibit the same behavior when her sons brought friends to the house and the noise rose to a level only adolescent boys can create.

With Mae's photo in my hand and Sendy on the phone, I started to communicate with Mae. The first thing Mae did after we said hello was to show me a mental picture of a racetrack, with its typical chaos and frenzy. Then she showed me a picture of her trainer—a tall, thin man wearing a leather cap and jacket—harshly pulling her from her crate and shoving her into a gated pen. She gave me a sense of her life as a race dog: confinement, running, then confinement again; lots of people making lots of noise; a gun firing. The more she told me, the more I felt her terror. I felt chills going up and down my back and marveled at how well she was doing with this kind of abuse as her legacy.

The leather jacket and cap that Sendy's brother wore, and the noise the boys made, reminded her of the track. She forgot that she was with new people in a whole other life, and her fear kicked in. She would flashback to the track, then cower, then pace, and then growl. This is classic post-traumatic stress.

Managing these situations required constant vigilance. I told Sendy that Mae didn't do well with surprises, and that she needed to know what was going to happen. I also coached Sendy on how to give Mae clear messages—making sure her thoughts and feelings were consistent and sending clear, sharp mental images along with her words. She decided to brief the boys on the strategy she would use with Mae, so they could do it, too.

They began by telling her when they would be returning home each day. They usually remembered to give her advance warning of anything unusual that was going to happen.

Sendy asked her brother to wear something other than his leather jacket and cap when he visited. This helped a lot. Mae was still wary, and Sendy's brother was still guarded, but the snarling and growling stopped.

When the boys' friends visited and Mae started showing signs of distress (whining, pacing, and panting usually preceded the growling and snarling), they would mentally tune in to her and silently remind her that she was with them now, in her forever home, and that she would never have to go back to the track. In spite of their initial skepticism, they found that Mae's agitation diminished whenever they followed this procedure.

Mae lived until she was fourteen, and never grew complacent with her new life.

It didn't matter to her that she didn't have a field to run in. She would run circles—first in one direction and then the other—in the flat area of her backyard; and when she tired of that, she chased her tail. She would sit on a hill and just observe, while other dogs in the neighborhood were barking at everything in sight. She only barked at squirrels and rabbits, which she would chase whenever the opportunity presented itself.

She loved when her dog friends came to visit—as long as they didn't stay overnight. She said she was very happy being an only dog.

And she loved stuffed animals. Teddy bears were her favorite, but rabbits, raccoons, and squirrels were acceptable replacements. One of the boys had a whole stash of stuffed critters. Mae would slip into his room, snatch one, and proceed to de-squeak it and bite off any plastic parts (which would sometimes show up during poop patrol). She always took tender care of them afterwards.

Mae found that the past doesn't have to predict the future. She learned to trust that humans can be kind and sensitive in spite of early experiences with those who were mean and cruel. Her ability to discern who would be safe unfolded throughout her life.

Mae's Message:

Sometimes things in our new life make us remember our old life. Remind us that the bad things will never happen to us again now that we're with you.

Jasper

"I Can't Do My Job."

Jasper, a longhaired mix of many cat breeds, had the face of a Maine Coon and the temperament of a wildcat. He was never very easy-going, according to his person Kara. However, things really changed after a move.

Jasper, another cat companion, and Kara moved into a different house with three other people and a dog. It was a little dog—a cream-colored lhasa apso—but it was still a D-O-G. Jasper had issues with the whole communal living/working situation.

The house was a hundred-year-old row house. Offices were on the first floor, and the living area was on the second and third floors.

Jasper took on the job of "protector" for Kara. He took up his post on the third step in the house, hissing and spitting at everyone—people and animals alike—who came near the staircase. At times, he even "went psycho"—with wild, wide eyes; tension; and general fear and anger. The only way to get him to settle down was to hold him until the tension released from his body.

Kara called me to try and understand what was happening to him.

As soon as we tuned in and I introduced myself, Jasper had a lot to say. He sent me the feelings of sadness and anger. He said he could see these people and didn't like them. (Most animals can see auras.) They were mean to Kara. She was pulling into herself. He wanted to be connected to her heart and her mind the way he always was. He couldn't do that when she was so inside herself, and that scared him. Kara acknowledged that she was withdrawing. The agreements the people reached before moving in together were not being followed, people spoke only a few words to each other, the work was not being shared, and there was no sense of community. It was not what she had counted on.

On further probing, Jasper said he had no claws to fight with if he had to protect himself or her. Unaware of the psychological trauma some cats experience with the procedure, Kara had had Jasper declawed when he was a kitten. Now she felt guilty that she added to his stress, and wondered if she made him so reactive.

I suggested that she apologize to him when they were in their quiet communing time together, and to let me know what happened. She called the next day to tell me that she apologized to him for her ignorance, and told him that if she could do it over again, she would never have taken his claws away. She said he noticeably relaxed and she sensed relief from him.

"The tension and anger that had been so much a part of him miraculously disappeared," according to Kara.

Jasper never relinquished his post on the steps, but his hissing significantly decreased. Once they moved out of that situation and returned to their original home, Jasper returned to his normal companion behavior—eating, sleeping, playing a little, sleeping, eating, taunting his "brother," sleeping. His "protector job" was complete.

Life went along comfortably for Jasper for several years. His retirement job was to alert Kara when anything upset his delicate sense of balance. He did this by pooping outside of his litter box.

When he was sixteen, his vet found a mass in his intestines. The only treatment was surgery, and it was questionable how successful it would be. I told him he had something in his belly (showing him a mental picture of a tumor), and told him we could either leave it alone and he might die from it, or the vet could remove it (showing him a picture of him sleeping and then waking up with a bandage around his middle and a pain in his belly). I got a clear **no** at the idea of surgery.

About a month later, he told us he could not recover; it was his time, and he wanted help.

Jasper wanted Kara to celebrate because this was exactly what he wanted. His work was finished, his purpose accomplished, and he had no fear of dying. He could last a little longer, but he didn't want Kara to see him weak and helpless.

Kara asked her vet to euthanize Jasper at home. He hated car rides as much as he feared the vet's office, and she wanted his transition to be as peaceful as possible. She called her friends who knew and loved Jasper, and arranged a celebration for the day his ashes were returned. They toasted Jasper, they toasted life in all its forms, and they reminisced about his "attitude."

I spoke with Jasper the next day. He said he liked all the people who were there. He loved having a party just for him. He liked that they said things about him that were true. He was glad they celebrated life. And he was pleased that he felt a direct connection with Kara and that she did the same with him. All in all, he was pleased with the festivities.

Jasper's Message:

When we suddenly behave differently, pay attention. It means that something is wrong. We don't do things to get even.

Chango

"I'm a Partner, Not a Pet."

Chango is a big, strong, and very determined springer spaniel mix. His attitude is that the best defense is a strong offense. Through the years, it has upset his person, Maya, a lot.

From the time he was a puppy, he would bark ferociously and lunge at any dog he saw. He would jump on people when they came to the house. He would snarl and growl at whomever was grooming him.

When he was harnessed in the car, he would try to break free before Maya was out of the car herself. He would whine and strain against the harness anytime she left him in the vehicle.

He was driving her crazy, and she was so upset with him that she seriously considered finding him another home.

After reading an article about animal communication, Maya searched for a practitioner in her area. It was her last resort.

In our telepathic phone conversation, Chango told me his job was to protect her, and he did this by forcefully telling other dogs to stay away from them. He was letting them know that he was there to protect his people and his house. He knew what he wanted and knew what he liked. He was confused by how upset Maya was. He was her partner. It distressed him when she yelled and gave him orders; he wanted her to either ask him to do something or tell him what she wanted.

Life is a serious matter for Chango. He has no use for playing; he considers it silly. He is keenly intelligent, and Maya described their training and agility lessons as exercises in "attempted perfection." She never found a way to make training a game, and Chango had no patience for repetition. She said he would become frustrated if he didn't get what he was being taught the first time he was introduced to it.

We developed a strategy to help Chango.

Maya practiced focusing her thoughts on what she wanted him to do instead of what she wanted him to stop doing. This was a challenge for her because she usually managed several things at once, and focus was not her strongest skill to begin with. Having to be aware of what she was thinking and making sure it was helping to move Chango toward what she wanted was a huge commitment. I also suggested that she be more consistent with him, telling him what she wanted and praising him when he did it, instead of being calm and tolerant until her last nerve was plucked, and then yelling at him.

We asked him to act a little differently when he went for a walk. We asked him to tune into Maya (or her husband, Chuck) and to use their reaction to help him learn what was safe and what was dangerous. If they were calm and happy, it was safe. If they were tense or fearful, it was dangerous. We asked him to bark a greeting to other dogs and then go back to calm alert. We asked him to wait calmly in the car until Maya completed her errands.

Chango worked very hard to please Maya, and, for a while, life was fairly peaceful and uneventful.

Then Maya became pregnant. She was worried how Chango would be with a baby taking up so much attention. Chango didn't know what a baby was. We told him it would mean another very small human in the house, who would need his protection. To prepare for this, Maya and Chuck played tapes of a baby crying, and practiced pushing a stroller with Chango walking by their side.

Once the baby was born, Chango became even more hyper-vigilant. He was always on duty, alerting them whenever the baby cried. He would only go for a walk if they all went, or if the baby and either Maya or Chuck were asleep.

He started to understand and tolerate playing a little more when they told him that human babies learn by playing.

As Maya understood why Chango does what he does, her appreciation for his complexity and his devotion to them increased. She apologized to him for being so impatient with him when he was a puppy, and for threatening to find him a new home.

A lot of what humans do still confuses Chango. He doesn't understand why Maya and Chuck are so tense (they bought a new house, Chuck had changed jobs, and Maya was working part time in addition to the new baby), and he occasionally still acts aggressively in reaction to their feelings. He settles down fairly quickly when they remind him it's just how humans are sometimes.

Chango and his people have developed mutual respect over the years.

In his early years, lots of people offered Maya advice on how to handle him. She tried and then rejected much of it because it didn't work or it didn't fit into her philosophy about how dogs should be treated.

"He has taught me to relax and trust my instincts instead of listening to every piece of advice every person offers," Maya said.

"He also taught me how important it is to let dogs—and people—be who they are. When I tried to make him act a certain way, he would balk. I couldn't make him into what I thought he should be. Our relationship changed when I let him be the serious protector that he is so naturally."

Maya is pleased to have learned this while her baby was so young. She regularly tells Chango what a marvelous teacher he has been for her.

Chango's Message:

Stop misjudging us. A lot of us want to be more than pets.

Tabasco, Albert, Toby, and Aster

"Know Who You Are."

Torri was one exasperated alpaca owner. She and her husband Dave had purchased four alpacas to join the menagerie on their farmette, with expectations that they would easily fit in—just as the dogs, cats, chickens, guinea pigs, ducks, and parrot did.

"When are they going to get nice?" she asked. "The owners of the alpaca farm told us they're tame, and all we have to do is halter them and walk them around and in no time they'll love us and want to be petted. They won't let us even touch them."

With this introduction, I asked her to send photos of Tabasco, Albert, Toby, and Aster. I also wanted to know how she and Dave came to acquire the frustrating foursome.

It turned out they bought the alpacas on an impulse. Torri wanted goats; Dave said no. Someone at his workplace suggested they look at alpacas. So, they visited an alpaca farm.

Torri said they were the most majestic creatures she had ever seen. She was fascinated with Tabasco because of his poise and confidence. They bought him and three companions on the spot.

The alpacas apparently knew these people were rank beginners, and set about indoctrinating them. It started with the alpacas backing off and spitting whenever either of them came near. It was followed by the alpacas' hanging around if Torri were outside reading or otherwise keeping to herself, and running off at any overture of engagement. They would occasionally eat out of a hand-held bowl, and then take off chasing each other or one of the other animals.

Initially, Torri and Dave felt anxious and insecure; and then, as the behavior continued, they became increasingly frustrated and aggravated. These alpacas were not acting like any of the other animals they surrounded themselves with, including the local skunk whom Torri fed on a regular basis.

That prompted them to call and find out why these animals were acting this way. Their study and coaching had not prepared them to be treated like servants.

Looking at their photos, and with Torri on the phone, I mentally called all four of them in. Tabasco was the only one who connected to me. I got the image of him standing in front of the other three, as if he were the spokes-alpaca for the group. I then got the thought that each alpaca has a role in the herd, and the others respect it. In this group, he is in charge; Albert keeps them light and laughing; Aster keeps them grounded and steady; and Toby makes sure they all cooperate.

Then Tabasco sent me a feeling of being disconnected, and a mental image of people (I couldn't make out who they were) talking and gesturing, and the thought "Why do humans pretend so much?" popped into my head. I asked what he meant, and the thought came in that "A lot of humans are not who they think they are." He continued by offering his opinion that alpacas see their human companions as caretakers, not equals. Alpacas think they are superior to any other animal they have met, including humans.

Tabasco was adamant that alpacas should never be compared to dogs and cats. "We are not domesticated," he thought to me. He sent me the feeling of being connected to the other alpacas, but not to Torri or Dave.

This information confirmed Torri's observation that these alpacas are independent (from humans, not each other), wise, and "always one step—or maybe two—ahead of us."

Tabasco also sent me a mental image of wearing a halter and accompanying Torri or Dave down the street, along with a feeling that can only be described as disgust. I asked him what he disliked about the halter, and the thought "submissive" came into my mind. Torri said she had already stopped haltering and walking them because she knew they hated it. Now she knew why.

Torri and Dave now describe the relationship as a "good roommate" situation: Say good morning, but don't expect hugs and kisses. These are not animals who will tolerate being coddled.

Torri and Dave say they are working on being clear and honest, making their thoughts, feelings, and actions congruent. It's the only way they will be able to have a relationship with the alpacas. The animals are keeping watch.

Tabasco told me they are learning to understand humans, but still feel more comfortable with other species.

Tabasco's—and the herd's—Message:

Don't try to make us into pets. You will not succeed. Know us. Know yourselves. That is the only way to be.

Rocky

"What Do You Want Me to Do?"

Rocky, a five-pound maltese, was six years old when his behavior radically changed. Always active and playful, he became increasingly lethargic, losing his balance, and getting exhausted with the slightest exertion. He could barely walk up the sidewalk, and his human mom, Arlene, would often have to carry him home.

Desperate and worried, she took him to the vet—four different ones, to be exact. None of them could find anything wrong. The mystery was compounded by the fact that Rocky would run around the clinic playing keep-away and entertaining the staff. The vets started talking MRI and exploratory surgery as Arlene continued to insist that something was wrong with him.

Knowing that conventional medicine was failing him, and intuitively sensing that he was dying, she began searching for alternative approaches. When she called me, she was grasping for a lifeline for Rocky and herself.

My first contact with Rocky was by phone. I had his photo, so I knew what he looked like. Arlene was on the other end of the receiver. I asked her to get as emotionally neutral as possible, then had her visualize Rocky and call him silently. Emotional neutrality on the owner's part is crucial for me. I want to be certain that I am connecting to each of them without the person's feelings overriding the animal's. When I tuned in, I felt sadness and anxiety from Rocky, and the thought "failure" popped into my head. I quickly learned from Arlene that Rocky had taken on a huge job, with no way of knowing that it was an impossible task.

Rocky's dad, Don, was in end-stage renal cancer. He had survived a series of medical crises through experimental procedures, but was getting weaker and wobblier with each treatment. In the way that many companion animals do, Rocky was taking on the symptoms in an effort to help Don get better.

I felt Rocky relax as I acknowledged that he had taken on a gigantic job. I also felt he needed immediate relief, and suggested trying flower essences with him. Flower essences are liquids made from plants and flowers, and hold their healing vibrations. They work subtly on the body's energy field to heal and restore harmony. They are simple to use: They can be added to a water bowl, or can be dropped directly into an animal's mouth or onto the fur and rubbed in.

Essences from the Green Hope Farm animal wellness line seemed most appropriate for Rocky, so I mixed a formula of anxiety, caretaker, grief and loss, and run and play for him. I also performed a distant healing, using a technique of creating harmony in the energy field which I learned from my healer friend, Deena Spear.

After Rocky was on the essences for about a week, Arlene reported that he seemed happier and was able to walk a little bit further.

Rocky said he was feeling better, but I still sensed an underlying anxiety. It was subtle, but I knew there was more to this story.

I asked Rocky how he spent his day with Don (visualizing the two of them sitting on the couch). Rocky admitted that Don was talking to him about dying, and how afraid he was for himself and for Arlene. Don said he was counting on Rocky to take care of her. Rocky had become Don's confidant, and didn't know what to do to fix the problem. The job was overwhelming him and crushing him. The only relief he got was when he was away from the environment, and the only time that happened was when he went to the vet.

Arlene was astonished that Rocky had taken on such a burden. She didn't know animals could—or would—do that. Yet it is far more common than most of us know. In their love for us, our companion animals will take on our symptoms to try and heal us—or to show us what we need to pay attention to.

It took several sessions, coupled with the flower essences, for Rocky to sort out his job.

I first explained to him that the way to really help Arlene and Don was for him to be healthy, playful, and a good friend. I sent him the feeling of joy, and the mental picture of him running around the house like he did before, playing catch with his favorite stuffed animal, and lying next to them as they watched television. I told him that trying to take on Don's pain and limitations would not make Don better. In fact, it was making things worse because his people were afraid he was really sick.

Rocky thought that being funny and getting them to laugh was not right because everyone was so sad.

I suggested to Arlene to start telling Rocky that she was sad and nervous because daddy was dying. She was handling it, and she and Rocky would be okay. She frequently reminded him about being a good companion, and that it was not necessary to take on daddy's dizziness and fatigue, or her anxiety.

Arlene also carefully told Don that Rocky was trying to save him. She told him it was crucial to Rocky's well-being that Don reinforce her message that Rocky was a great companion, and that was the most important job he could take on.

Then Arlene and Don started talking to each other about their life together, their fears, their sadness, and their anger and the unfairness of it all.

Over a three-week period, Rocky "miraculously" changed. Arlene reported that he perked up, played with his toys, entertained them, and insisted on long walks.

So Rocky became both a catalyst for and facilitator of a level of human intimacy that his people never imagined possible. Arlene said it deepened their relationships—both hers and Don's and hers and Rocky's—in ways she never expected.

To this day, Arlene continues a practice she began then. She tells Rocky what is going to happen in their lives. She tells him her feelings, and thanks him for listening and for just being there. She says that whenever she tries to hide her feelings and tell Rocky something she thinks will make him feel better, he acts confused and anxious. She says Rocky has been a phenomenal teacher about intimacy in relationships.

She and Rocky shared their grief after Don died. They did better than she expected. Rocky now spent a good part of the day by himself while she worked. She made sure to tell him how the day would go, and what time she would be home, and he settled into the new routine.

About a year later, Arlene's friends dragged her to a singles event—and she met a new man. He passed muster with Rocky, which was one of her requirements if she were to get into a serious relationship with anyone. They got married the following year.

Rocky feels secure in his relationship with her. The mutual love and appreciation they hold for each other, and how Arlene has communicated it to him, have left Rocky very sure of his position as mom's #1 boy.

Rocky's Message:

When you are sick or sad, we want to make you better. We feel bad if we can't. Tell us what you want us to do.

In Conclusion ...

Here are sixteen remarkably unique, yet remarkably ordinary, animal companions. Each agreed to have his/her story told, and all of them were pleased to be invited to offer their wisdom to us.

Our animal companions want to be seen and heard as much as we do. Some are grateful when we listen, and others demand it. Often, giving them the opportunity to tell us what they want is enough to stop their niggling misdeeds.

Their messages are simple and clear.

Intentional deceit is not a part of their makeup. When all is well, all is well. When all is not well, they get our attention in whatever way they can.

They do what is natural to them, yet willingly temper their feral instincts to live with us. Those who make the decision for domestication like living with us, but will always be cat or dog or bird or horse. They want us to remember that they are not people in fur or feathers.

Welcome to my world. It's a wonderful place to live.

To contact Terri Diener

terri@petspeak.com
www.petspeak.com

410-486-4849

Copies of The Pets Speak may be ordered directly from Terri

For information on sponsoring a workshop on animal communication, contact Terri by e-mail or phone

Printed in the United States
124894LV00001B/6/A

9 781595 941176